REPORT OF THE INQUIRY INTO CABLE EXPANSION AND BROADCASTING POLICY

Chairman: LORD HUNT OF TANWORTH

*Presented to Parliament
by the Secretary of State for the Home Department
by Command of Her Majesty
October 1982*

LONDON
HER MAJESTY'S STATIONERY OFFICE
£4.40 net

Cmnd 8679

Note: The estimated cost of preparing this Report is £47,388 of which £5,418 represents the estimated cost of printing and publication.

ISBN 0 10 186790 5

MEMBERS OF THE INQUIRY

The Lord Hunt of Tanworth, GCB

Sir Maurice Hodgson

Professor James Ring

Secretary: Mr J C Davey

Assistant Secretary: Mr P R C Storr

To The Rt Hon William Whitelaw, CH, MC, MP, Secretary of State for the Home Department.

You announced on 6 April 1982 that you had appointed us to undertake an inquiry into the broadcasting policy aspects of the expansion of cable in accordance with terms of reference which you had set out in a statement to the House of Commons on 22 March.

We have completed our inquiry and now have the honour to submit our report.

HUNT OF TANWORTH, *Chairman*

MAURICE HODGSON

J. RING

JON DAVEY, *Secretary*

PETER STORR, *Assistant Secretary*

28 September 1982

CONTENTS

		Page
CHAPTER 1:	**The Inquiry's Task**	1
CHAPTER 2:	**The Broad Perspective**	2
CHAPTER 3:	**Functions in Cable and the Case for a Franchise System**	6
CHAPTER 4:	**Ownership of Cable Operators**	9
CHAPTER 5:	**Cable's Sources of Income**	11
	Rental	11
	Subscription	11
	Advertising	12
	Pay-per-view	16
CHAPTER 6:	**Cable Programme Services**	17
	Number of channels	17
	The "must-carry" rule for public service broadcasting	18
	Carrying of other United Kingdom broadcasting services	20
	Carrying of foreign broadcasting services	20
	Range and balance of programmes	21
	Exclusive rights	22
	Local access	23
	Taste and decency	24
	Impartiality	25
	Foreign material	25
	The cinema	26
	Copyright	27
CHAPTER 7:	**Oversight**	28
	The nature of the task	28
	The purpose of franchising	28
	Subsequent oversight	30
	Local or central franchising?	30
	The cable authority	31
	Sanctions	32
CHAPTER 8:	**Summary of Conclusions and Recommendations**	34
APPENDIX A:	Consultation document issued by the Inquiry on 7 April 1982	40
APPENDIX B:	List of organisations and individuals who made written submissions or with whom discussions were held	43

CHAPTER 1

THE INQUIRY'S TASK

1. The Home Secretary announced on 22 March 1982 that this Inquiry was to be established with the following terms of reference:—

"To take as its frame of reference the Government's wish to secure the benefits for the United Kingdom which cable technology can offer and its willingness to consider an expansion of cable systems which would permit cable to carry a wider range of entertainment and other services (including when available services of direct broadcasting by satellite), but in a way consistent with the wider public interest, in particular the safeguarding of public service broadcasting; to consider the questions affecting broadcasting policy which would arise from such an expansion, including in particular the supervisory framework; and to make recommendations by 30 September 1982."

2. Our timetable was accordingly tight and some representations were received that more time should have been allowed. It is right therefore to record our view that the time we were given was sufficient to allow us properly to consider the issues raised by our terms of reference. It was long enough, following the issue of our consultation document on 7 April (Appendix A), to allow no fewer than 189 organisations and individuals to submit written evidence to us, all of which we carefully considered; and for us in many cases to take oral evidence from them as well. The Inquiry held 32 meetings. In addition, visits were made to the United States and Canada, where we had useful discussions with many people involved in one way or another with cable and broadcasting, and to two of the subscription television pilot schemes in this country. The full list of those who made written submissions or with whom discussions were held is at Appendix B and our thanks are due to all of them.

3. We also wish to record our thanks to our own Secretariat. Mr Jon Davey, the Secretary to the Inquiry, has been unfailingly helpful and resourceful. The fact that we had only six months in which to undertake the Inquiry, combined with the very large amount of evidence submitted to us, imposed a heavy burden of work on him, on our Assistant Secretary Mr Peter Storr and on Miss Juanita Rodney. They never flagged and we are very grateful to them.

CHAPTER 2

THE BROAD PERSPECTIVE

4. The decisions which must be taken in relation to cable expansion are of great significance. There are no modern cable systems in this country. We have some ageing narrowband systems, which do no more than relay public service broadcasts except in the case of a handful of pilot schemes of subscription television which add a single channel consisting almost entirely of films. The new wideband cable systems can provide not only a very large number of channels but also channels with two-way communication capability ("interactive") allowing information to pass in both directions. For example, cable operators can put questions to viewers through the television set and get their response; viewers can have fire or burglar alarms which can alert the emergency services; and, more fundamentally, interactive channels can provide rapid communications between businesses and between them and their customers. At its best therefore cable can help both business and the individual by providing new methods of working, buying and selling direct from the home; increased facilities for education and training in the home; services like electronic mail and tele-banking; and a greatly increased and enriched choice of home entertainment. At its worst, however, it could lead to a waste of resources, risks to privacy and a lowering of the quality of broadcasting. It is therefore crucially important that the right decisions are taken.

5. This Inquiry was not concerned with the technology to be used for cabling the country, for example, whether coaxial or fibre optic cable should be used. Nor were we concerned with the provision of the interactive services for business and the consumer. These matters are under separate study by Government departments. There is however one general point which needs emphasising. It seems to be generally agreed that investment in cable television for entertainment purposes will be the necessary base to which the interactive services of economic benefit to business and the individual will be added. It is therefore very important that decisions about cable television should be taken with this fact in mind and that for example the award of franchises should positively encourage the development of the interactive services.

6. While television will be only part of a cable system it will be a very important part. How should it be regarded? And why, to quote our terms of reference, is it necessary to consider "the safeguarding of public service broadcasting"? What indeed distinguishes the latter?

7. The ability to receive BBC and Independent Television throughout the country at no charge after the payment of the licence fee has clearly been one important characteristic of public service broadcasting. Equally important however are the quality and wide range of the programming. Reith's aim was "the best of everything for everyone": and both BBC and ITV are publicly accountable not only to maintain high programme standards with a large amount of original material but also to put out a balanced selection of information, entertainment and education, with something of interest to all sections of society and to all ages. They have a formal obligation to do this and it has a

significant effect on their programming schedules. Furthermore, while the BBC and ITV naturally want to reach large audiences, they are shielded from direct commercial pressures on their programming. It would be a serious loss if programming of such quality and diversity were eroded either through a general lowering of standards ("the bad driving out the good") or if quality programmes tended to become available only to people who could pay extra for them and who lived in those parts of the country which had been cabled. If the recommendations in our report are accepted, we do not think this is likely to happen. Furthermore we believe that recent developments both in multi-channel cable technology and indeed in public service broadcasting itself justify a reassessment of the arguments against subscription television which the Pilkington[1] and Annan[2] reports found convincing. We are however satisfied that some limited safeguards against damage to public service broadcasting will be necessary in the interest of the large sections of the community who, through choice or necessity, will remain dependent on it.

8. We believe that multi-channel cable television should be seen as supplementary, and not as an alternative or rival, to public service broadcasting. In Chapter 6 we make it clear that all new cable systems must carry all free BBC and Independent Television services as part of their basic programme package both so that no one should be deprived of them by a decision to connect to a cable system and also to preserve the habit of national viewing alongside the new channels which will cater more for special interests and also be more locally orientated. Apart from carrying these basic services, cable television should be different and should greatly widen the viewer's choice by providing many additional channels for those people ready to pay more than their basic television licence fee. Because financing and other constraints will limit the geographical area of any one cable system it should be responsive to local demand. There will therefore be no such thing as a "typical" cable television package. Assuming however a system with more than 30 channels most packages would normally comprise three distinct sections: first, the "must-carry" broadcast services referred to above; second, some interactive services of benefit to business and the consumer; and third, a larger range of channels providing such programmes as sport, films, arts, continuous news, education (including the Open University at convenient times), children's features, hobbies, health, games and locally derived programmes whether they be community affairs, ethnic, local arts or channels providing access for local people to make their own programmes.

9. Cable television is therefore all about widening the viewer's choice. It should be innovative, experimental and sensitive to local feeling. It cannot be run as though it was another branch of public service broadcasting providing a balanced service for the country as a whole. The multiplicity of cable systems all offering many channels would in any case make this sort of control impossible without a large bureaucracy which would stultify the initiative and diversity that will be both inherent and desirable in a cable system. The whole approach needs to be different, with the maximum incentive and encouragement for responsible initiative.

---

[1]Report of the Committee on Broadcasting, Cmnd 1753.
[2]Report of the Committee on the Future of Broadcasting, Cmnd 6753.

10. There is no reason why we should not have a high standard of privately financed cable television, just as we enjoy a high standard of public service broadcasting. Nor should we look too much to the experience of other countries where cable television developed for quite different reasons: for example, the inadequacy of "off-air" reception and the lower standards of public service broadcasting in the United States, or the desire to bring in programmes from neighbouring countries in parts of Europe. A high standard will not however necessarily be maintained of its own accord, particularly given the understandable wish to see an early return on the considerable investment that will be needed. In the early stages there will also be an inevitable shortage of good material made in this country; and while no doubt there are many creative writers and producers who will seek outlets for their ideas in the world of cable television it must be remembered that the making of original programmes for television is much more expensive than, for example, panel games, the showing of old films, etc.

11. We have carefully considered the thesis that a cable television system, as a result of its diversity, can be seen as just another branch of publishing, subject only to the laws on such matters as defamation, sedition and obscenity. We believe this approach might become acceptable when the whole country is cabled, if there is then genuinely the same sort of choice as is found in a bookshop, and if there is a general consensus about what has, and what has not, got a place on cable television. In the early days the capacity of some systems will still be limited: there will be a need for diversity and innovation to be matched by some consistency over access and standards: and the weight of evidence submitted to us strongly suggests that most people still see the carrying of programmes into the home as different in kind from the act of going out to buy a book or a magazine. A book is read alone, whereas much television is watched by the family as a whole. Furthermore we agree with the Williams Committee[1] that film with all its associated techniques of close-up and special effects is a uniquely powerful instrument, and we regard video material as within the same category. We deal later with whether a more permissive attitude would be appropriate if electronic "locks" are fitted, which allow the subscriber to restrict the reception of some programmes.

12. Cable television must therefore operate within certain, albeit liberal, ground rules. Some of these can be settled at the outset and we make recommendations accordingly. We are however satisfied that there will be a need for some continuing oversight. We do not think that "self-regulation" by the cable industry would be generally acceptable at a time when cable is establishing itself and when, for example, there are clearly differing views within the industry itself about the showing of programmes which could be offensive to many people. However, we emphasise the word "oversight" rather than regulation. The latter implies imposing detailed programming rules across the country whereas the whole idea of cable is that it should supplement public service broadcasting and be responsive to local demand. The concept should be one of cable operators accountable to observe a few general guidelines and provide the service offered when given their franchises rather than one of a central body regulating in detail how they go about their business.

---

[1]Report of the Committee on Obscenity and Film Censorship, Cmnd 7772.

13. There is a further reason why some degree of oversight seems inevitable. While we have confidence in the recommendations we make for the introduction of modern cable television systems we are far from confident that we can accurately foresee future developments. Cable television is in many ways a leap into the dark precisely because it should be experimental, diverse and attuned to local needs. At this stage it is impossible to know what sort of channels people will most want. General oversight, as opposed to detailed regulation, will provide the opportunity to respond flexibly as the industry develops in ways which are impossible to forecast now.

14. This chapter has set out our basic approach to cable television. The following chapters describe in greater detail the issues which will arise and contain our recommendations on the decisions to be taken at the outset and on the arrangements for subsequent oversight. We begin with the way in which cable systems are likely to be established.

CHAPTER 3

FUNCTIONS IN CABLE
AND THE CASE FOR A FRANCHISE SYSTEM

15. We found it helpful to identify four functions which, although interrelated, can be seen as distinct and requiring different kinds of expertise. They are as follows:—

(a) *The cable provider*—the installer and owner of the physical infrastructure.

(b) *The cable operator*—the manager of a local cable system who puts together a package of cable services to sell to subscribers in his area, from whom he collects revenue. For this purpose he will draw on various sources in order to assemble the most attractive service with which to attract the greatest number of subscribers.

(c) *The programme or service provider*—the assembler of programmes into channels or segments of channels for sale to cable operators, consisting for example of cinema films, news, sport, education, children's programmes, etc; or the provider to the cable operator of other services (local information, home security, teleshopping, etc).

(d) *The programme maker*—the ultimate source of the programme material which is assembled into channels by the programme providers.

16. Our conclusion is that the key figure of these four will be the cable operator. The evidence we have received strongly suggests that the main commercial motivation lies in selling services to the public. It is therefore probably the cable operator who will take the initiative in putting together proposals for new cable systems. An alternative view that the cable provider will take the initiative seems less likely because he is more remote from the eventual sale of services. It is significant that the minority of witnesses who saw the cable provider as the key figure were not themselves proposing to install cable systems. Those who wish to participate in the cabling of the United Kingdom are primarily cable operators—and we class those companies who also intend to install their own systems as first and foremost cable operators, because their wish to install cable is secondary to their main purpose of operating systems once they are in existence.

17. Our perception of the cable operator as the key figure is in any case not dependent entirely on our view that he will normally be the initiator of cable expansion. Even if a cable provider were to take the initiative, the key figure for our purposes, which chiefly concern the supervisory framework for entertainment and other services to the public, would still be the cable operator, since it will be he who will be answerable to the viewer for the service he provides. This fact is central to our approach to a number of the questions we have to consider.

18. In our view, therefore, cable systems will be installed because cable operators take steps to get them installed; but what are those steps to be? The Cable Television Association advocated continuation of the procedure presently applied to the existing narrowband cable systems which provide a very

limited range of services largely of a relay nature. Under this a cable operator obtains from the Home Office a licence under the Wireless Telegraphy Act 1949 (which permits him to receive wireless signals for the purpose of distributing them by cable) and another under the Post Office Act 1969 (which permits him to operate a programme distribution service) and then negotiates wayleaves (which permit him to lay cables in the street etc) with the relevant local authorities. We felt unable to recommend this procedure for multi-channel cable systems. We are clear that there are strong grounds for a formal system in which a franchise is granted for the operation of a cable system in a particular area. The principal reasons for taking this view are:

(i) The cable operator will have an effective local monopoly as a result of the high cost of installing and operating a cable system and of the need to achieve high customer penetration for financial viability. It is desirable that this monopoly should be conferred explicitly in a way which secures for the customer the best possible service rather than that it should happen *de facto* by licensing on a first come first served basis.

(ii) Franchising would provide the best way of facilitating the submission of competitive applications and the fairest mechanism for selecting the most desirable operator having regard to the channel capacity of the proposed system, the quality and variety of programming and the provision of other services, the degree of local access and the geographical area to be cabled.

(iii) Having negotiated the provision of the best system with the cable operator, a franchising authority would be the most appropriate body to oversee his performance in relation to the proposals put forward when the franchise was granted.

(iv) Without a formal franchising system, there is a risk that the grant of wayleaves by local authorities would be turned into an informal system of franchising with the criteria and conditions varying from authority to authority, some of which might seek to take powers which are unnecessary or inappropriate.

The form and nature of a franchise system, together with related aspects of cable oversight, is discussed in Chapter 7.

19. We do not believe it is necessary for the programme provider or the programme maker to be licensed; and given the large number of local channels likely to exist it would be very difficult to do this anyway. Moreover, licensing of programme providers would seem inappropriate if cable operators are allowed to relay services from overseas—and given the almost inevitable inclusion of foreign programmes in the services assembled by British programme providers it would be even more inappropriate to license programme makers. It will be the cable operator who determines what services are carried on his system and except for those channels for which a United Kingdom broadcasting authority takes responsibility it seems simplest to hold the cable operator responsible for the programmes and services he distributes. Programme providers will then only succeed if they produce material which the operator is able to use.

20. The question remains whether the cable provider should be licensed, for example in order to ensure that he complies with any necessary technical standards which might emerge from the studies on these matters now being undertaken by Government departments. This is not primarily a matter for us. Separate licensing could perhaps be avoided if the onus were placed on a cable operator granted a franchise to ensure that any system he operated met certain specifications.

21. In the evidence submitted to us and in our own deliberations a good deal of attention was directed towards whether, to avoid undue monopoly, there should be any enforced separation of ownership between the four roles we have identified or whether two or more could be undertaken by the same organisation. We now consider this point in the light of the key role which we see for the cable operator.

22. First, the relationship between the operator and the cable provider. If the cable system throughout the country were to be laid by a single body (whether British Telecom or a new authority) acting as a national common carrier which would then rent its lines to cable operators, there would of course be no need to consider separation between the provider and the operator since it would already exist. Although not within our terms of reference, the national common carrier model appears unlikely as it is inconsistent with the Government's policy on competition and its expressed view that cabling should not make significant demands on public expenditure. If on the other hand the cable infrastructure is to be financed privately (with the cabling of different systems being done by more than one body not excluding British Telecom) there will be a need for close co-operation between those raising the money for it and those who are to market the subsequent services, and in some cases we would expect the same company to wish to undertake both functions. Subject to one proviso, we see no reason to insist that these functions must be carried out by separate companies. This proviso is that in cases where the cable operator was also the cable provider a condition of the franchise given to him as operator would be that, to ensure continuity of service to his customers, he should sell or lease his infrastructure on a predetermined basis to another operator if he was deprived of his franchise.

23. We now turn to the relationship between the cable operator and the programme provider. Again we see no reason why the operator should not also provide programmes if he wishes to do so especially as these would clearly be only a small part of his programme package. It has been suggested that ownership links between cable operators and programme providers in the United States have in some cases restricted the choice for subscribers because cable operators who are associated with particular programme providers are less likely to carry programmes offered by others. We do not think there will be real risks from such vertical integration in this country in the foreseeable future. Indeed the problem is more likely to be one of a shortage of good material rather than good material not finding an outlet. However, in awarding a franchise attention should be paid to the programme package proposed to ensure that diversity of programme sources is encouraged and any undesirable monopoly avoided, and there should be an expectation that some channels should be available for leased use by persons having no connection with the cable operator.

CHAPTER 4

OWNERSHIP OF CABLE OPERATORS

24. A cable operator will be concerned with providing information and education, as well as entertainment, to his customers and he will also have an effective monopoly in his area. We are therefore satisfied that it is right that there should be some restrictions on ownership of companies which operate cable systems.

25. We believe that the ownership of cable operating companies providing monopoly services into the home should be free from any kind of political or ideological bias. We would exclude both central and local government, and also any political party or organisation, from direct participation in the ownership of companies operating cable systems. We think the same rule should apply to religious bodies.

26. The next question is whether there should be restrictions on foreign ownership, or ownership by the press or radio or television companies. In these cases we distinguish between a majority and a minority interest. We see no need to prohibit companies in these categories from participation in the ownership of cable operating companies. There is a great deal of experience of operating modern cable systems in other countries which is not available here. So far as the press is concerned, we see a significant role for cable systems in facilitating communication in local communities and we welcomed the evidence given to us that the press would like to contribute to these new developments. The experience of the television and local radio programme companies would also be valuable. But while these are all useful sources of expertise which could have a stake in a cable operation, we do not think any individual company of this kind should have a controlling interest. It would seem unacceptable for cable systems to be under foreign control, and control by another media interest could lead to undesirable monopoly power.

27. Apart from this, we see no need to suggest any other restrictions on the ownership of companies operating cable systems. It is likely that the resources needed to install and run future cable systems will bring together a variety of interests in consortia which will apply for local cable franchises. It is desirable that, because cable systems will be essentially local, there should usually be some local participation in the equity. However, we see no need to rule out the possibility of a single company, such as one which already owns and operates cable systems in this country, being allowed to proceed alone if it has the resources to do so.

28. We have already proposed that there should be some kind of franchising procedure, and it seems to us that it is within these arrangements that any other considerations relating to possible monopoly interests in cable operating companies can be taken into account. In the light of local circumstances, the franchising authority could decide if a particular level of cable ownership or involvement seemed undesirable.

29. We see little danger of a monopoly arising from excessive ownership of cable franchises throughout the country, as distinct from the ownership of individual systems. Although much of the evidence we received stressed that it would be undesirable for one company to own a substantial number of the cable operations in this country, it was widely thought that this was unlikely to happen. The cost of cabling the country will be such that no one company is likely to be able to undertake more than a relatively small share of the task. We do not recommend specific restrictions on the scale of operation of particular companies, but the franchising process could again monitor the possible danger of national monopoly. We deal in Chapter 7 with the kind of franchising arrangements required, but any question of maintaining oversight of the pattern of ownership throughout the country points to the idea of franchising being undertaken on a national rather than a local basis.

30. We see no reason to impose any restrictions whatever on the ownership of programme makers, cable providers or programme providers, apart from that in the last sentence of paragraph 76.

CHAPTER 5

CABLE'S SOURCES OF INCOME

Rental

31. The ageing narrowband cable systems in the United Kingdom at present charge those who use them a small monthly rental – usually not much more than one pound or so. This is now often insufficient to pay the cable system's overheads but all the customer is getting is the relay of off-air broadcast services, for which the cable operator pays nothing. A small incentive to subscribe to cable at present is that the cable operator, by using a superior aerial to that which most householders would consider installing, is often able to relay one or two out-of-area ITV services in addition to the local service, and this will at times give the subscriber a wider choice of viewing.

32. The basic rental charge for a cable system of the future containing 30 or more channels is bound to be more than a cable subscriber pays at present, since the operator will incur much greater costs, some of which will have to be recovered in this way. The customer will however get more for it. In addition to the existing public broadcasting services there may well be a wider range of out-of-area ITV services and foreign broadcasting services which cannot readily be received by an individual aerial; there will be services of direct broadcasting by satellite (DBS) which could well be cheaper to obtain by connecting to a cable system than by installing an individual receiving dish (though one BBC channel will be by subscription); there could be local information services and community programmes produced at modest cost; and there will probably be some interactive channels which will provide information services or facilities for home shopping, etc. However, if the basic charge is kept at a level likely to encourage the rapid expansion of cable, the cable operator's revenue is unlikely to cover very much more in the way of additional programme material unless the latter is financed by subscription or by advertising.

Subscription

33. Many people would be prepared to pay an additional charge to receive extra programme services of their choice. The present pilot schemes of subscription television involve cable subscribers paying an extra subscription for a channel which provides a programme of recent feature films; the indications are that the idea of paying to see in the home newer films than are being shown on BBC or ITV is attractive. Subscription for particular additional channels is a natural and, by the evidence we received, a generally acceptable way of financing additional programme services on cable. It is a new source of revenue and would not divert revenue from the BBC, which (until it starts its DBS subscription service) draws its income from television licence fees, or from ITV which is financed solely by advertising. We are however satisfied that the number of additional channels for which most people are prepared to pay extra is limited.

34. Admittedly, subscription rather than advertising has been the main source of financing for cable in the United States hitherto, though this is changing. But we think this is a respect in which American experience may be a

misleading guide to developments in the United Kingdom. One of the reasons for the success of subscription television in the United States is the attraction of getting away from the intrusiveness of broadcast television advertising. There is twice as much advertising per hour on the broadcast networks in America as in the United Kingdom and the commercial breaks are more frequent. In this country, the lower level of ITV advertising is much more acceptable to the viewer who already has a choice of watching television with or without advertising. For most people the presence of advertising does not affect the decision whether or not to watch a particular programme. Thus the motivation to pay for subscription television may be weaker here than in the United States, and as a consequence subscriptions may be a less buoyant source of finance and insufficient on their own to finance cable expansion. Equally so, there may not be the same antipathy to advertising in cable services provided it does not become too intrusive. It is also relevant that the programmes for which Americans are paying through subscription cable services are largely cinema films. Much of the wider range of American cable services is contained in the many channels which are provided nationally and offered as part of the basic cable package supported by advertising rather than by special subscription. These include the Alpha arts channel, the 24 hour news channels (like Cable News Network), and the sports channel (ESPN).

35. Accordingly we recommend that subscription for particular additional channels should be allowed: but we think it unlikely that significant cable expansion in the United Kingdom will be able to take place solely on the basis of rental and subscription.

Advertising
36. Advertising on cable raises issues at the heart of our terms of reference, since we were asked to concern ourselves with the safeguarding of public service broadcasting. The independent half of public service broadcasting is entirely dependent on advertising revenue and might be damaged if part of its income was drained away through the establishment of services which were in direct competition for the same source of revenue. It is also arguable that a deterioration in ITV programmes would have a knock-on effect on the BBC.

37. The questions we have considered are whether and to what extent the total amount of advertising is likely to increase if new opportunities to advertise on cable are offered; what kind of advertising is likely to be attracted to cable; whether and to what extent advertising will be taken away from existing media; which forms of media are most likely to suffer if this happens; and what the effects are likely to be.

38. Total advertising expenditure in this country in 1980 represented 1·32 per cent of gross national product. This is higher than in most countries in Europe but lower than in the United States, where the equivalent proportion is 1·6 per cent. It is also lower than has occurred in the past in the United Kingdom—in 1960 it was 1·43 per cent—but it has been growing steadily over the last six years. If advertising's proportion of GNP returned to its previous highest level in this country it would represent additional advertising expenditure at 1980 prices of about £200 million. If advertising expenditure in the

United Kingdom were to rise in relation to GNP to the level in the United States the increase in expenditure at 1980 prices would be more like £500 million. Moreover, advertising expenditure would be expected to increase with the growth in GNP, and this could result in an additional several hundred million pounds (depending on the rate of growth of GNP) being spent on advertising by 1990. Although it has been argued that the amount of money available for advertising is determined by commercial judgement irrespective of the number of media available, we find it difficult to believe that advertising revenue is fixed at its present level in real terms and that advertising on cable would necessarily mean the same size cake being sliced more thinly.

39. It is noteworthy that the fastest growth in total advertising expenditure in the United Kingdom took place in the late 1950s. This was the period during which television advertising became available for the first time and although by the end of the five years up to 1960 television had secured 22 per cent of total advertising expenditure, with the share taken by the press falling as a result from 87·7 per cent to 70·9 per cent of the market, the growth in total advertising over that period was such that expenditure on press advertising actually increased in real terms. Fears about the loss of advertising revenue to the press following the introduction of television advertising turned out to be unfounded.

40. There will already be a significant increase in the availability of television advertising time, and hence in the competition for revenue when Channel 4 starts broadcasting in November this year. One suggestion put to us was that even assuming there is some elasticity in advertising expenditure, this will be pre-empted by Channel 4, which hopes to raise over £100 million a year through advertising revenue.

41. This was not a universal view. The Institute of Practitioners in Advertising thought that cable advertising would not erode the advertisement base of Independent Television but could by 1995 contribute £120 million a year (at 1980 prices) towards the costs of cable services. They also forecast that, after allowing for this, spending on Independent Television advertising could rise from £692 million in 1980 to £800 million in 1995, and that total annual advertising expenditure would increase over the same period from £2,562 million to £2,900 million. They emphasised that fears which had been expressed in the past about the effect of new media advertising on the viability of established media had in all cases proved to be exaggerated.

42. The effect on existing media will partly depend on the type of advertising which is attracted to cable. The Cable Television Association suggested to us that the local nature of cable systems would make them unsuitable for the kind of advertising which is carried by Independent Television. This was a premise that we do not wholly accept. We think it unlikely that cable systems could become sufficiently attractive to the public solely on the basis of local programming. It is likely that, as in the United States, national services will develop, fed to local cable systems by means of trunk cable networks or microwave links or satellites, and these could be financed by a combination of national and local advertising. Some cable advertising may therefore be of a national character. Even on countrywide services, however, some advertising

could be placed locally by national advertisers who wished to support the test-marketing of new products in certain areas, for which purpose the present ITV regions are sometimes too large. It might also be that some national services would be specialist in nature and therefore attractive to advertisers interested in reaching a target audience and for whom television advertising would normally be too expensive. The trend in advertising is very much towards identifying the right audience instead of simply looking for the largest audience and it was suggested to us that specialised channels would be fertile ground for more specialised advertising. Advertising of this nature would be unlikely to detract from the revenue of Independent Television and could well represent entirely new spending.

43. We agree, however, that one of the strengths of cable will be its local nature and its ability to attract local advertising. For the first time, television will become a medium which will be available to local shops and services, and advertising by them will increase the overall amount of television advertising and represent new spending. Some of it, however, could be drawn away from the local press and this could happen particularly if cable were able to offer a service of classified advertisements, of jobs or of property for example, as happens in the United States. Cable channels would be particularly well-suited to offer new forms of classified advertising. The viewer could select at his set only the advertisements he wished to view from a very much larger number which were available. The reaction of the Newspaper Society, representing the provincial and local press, towards these possibilities was a positive one. They did not oppose advertising on cable, provided they were allowed to participate in cable systems. As we said in paragraph 26, we see no reason why the press should not have a stake in cable operation provided it did not amount to control.

44. To some extent the possibility of advertising on cable television could also threaten the income of Independent Local Radio. This was a matter about which the radio contractors expressed concern to us; in particular, it was the local element in cable advertising which they thought might detract from that part of ILR revenue (about half) which came from local advertising, and they proposed that cable services should be financed solely by subscription. It is difficult to assess the validity of these fears but we noted that the forecast of the Institute of Practitioners in Advertising was that ILR advertising revenue would grow in real terms notwithstanding the competition from cable, and we do not think that the risk is sufficient ground for banning advertising on cable.

45. Finally, in considering the effects on other media, we must record that fears were expressed to us by the cinema exhibitors, whose comparatively small advertising revenue is nevertheless frequently crucial to their viability. The kind of advertising shown in local cinemas could well be what cable systems would expect to carry in the future and we recognise the fear of the cinemas that some of this might be lost to them. However, their main concern was not with advertising but with the showing of films on cable, which we discuss in paragraphs 78–81.

46. Our terms of reference required us to take the effect of cable on broadcasting (rather than on the press or cinemas) as our main concern. Misgivings were expressed to us that if ITV or ILR suffered a large loss of revenue, they would be less able to meet their present obligations and would feel the need to compete for audiences through programming of a more popular character to the detriment of the public service concept. We should make it clear however that these misgivings did not lead the Independent Television Companies Association or their individual members to suggest to us that advertising on cable should be prohibited, and they were by and large content to face a certain amount of competition from this source (especially as they are expecting some compensation in the form of additional markets in which to be able to sell programmes).

47. All forecasts of additional advertising revenue and the proportion which will be taken by cable are hedged around with uncertainties: and these uncertainties are greater at a time of recession. We have however not been convinced that the likely level of advertising on cable would damage ITV or ILR to any significant extent in the short or the medium term; and both the advertiser and the consumer should benefit from the increased competition and the flexibility that cable can offer as an advertising medium. Furthermore, as we have already said (paragraph 35) we consider it very unlikely that the full potentiality of cable expansion could successfully be achieved without advertising revenue. We also believe that advertising on cable will have some positive benefits. Accordingly we recommend that it should be allowed.

48. Having reached that view, we considered the case for any restrictions on cable advertising, whether from the point of view of amount or as to nature. The amount of advertising on broadcast television is restricted, to an average of six minutes an hour and a maximum of seven minutes in any one hour, but it does not follow that advertising on cable should be subject to the same restrictions. The limitation on broadcast television is primarily to protect the viewer from excessive advertising, as part of the concept of broadcasting as a public service. Cable, however, would not be a public service in that sense and it would have to provide a service that was sufficiently attractive for the public to buy. Advertising that was so intrusive that it put the public off would be counter-productive. On the other hand, cable could play a useful role in providing a service of specialised advertising. The emergence of longer and perhaps more informative advertising could assist consumers to make product choices; and we have already referred to the possibility of a channel or channels devoted to classified advertising. Recruitment advertising in particular, it was suggested to us, could be very successful on cable. Another example would be a channel devoted to home shopping where, in effect, a television channel would become the equivalent of a mail-order catalogue. We see no reason why cable should not be allowed to devote the whole or any part of a channel to advertising in this way. If the intention is that some channels might consist of nothing but advertising, it becomes rather more difficult to restrict the amount of advertising which might be allowed on other channels. On balance, we believe that there are insufficient reasons for seeking to lay down defined limits at the outset. However, we have earlier in our report stressed the need to adopt a flexible approach to the future of cable and this is an area where, after some years of experience, it could become necessary to impose restrictions.

49. As to the nature of advertising, there is a clear need for observance of a code of practice, and we do not think that advertising on cable should adopt different standards from those which now apply to ITV in relation to products whose advertising is prohibited or the methods by which goods or services are advertised. We do, however, consider that there is scope for sponsorship of programmes on cable provided that certain clearly defined rules, including the separation of advertisements from editorial matter, are observed. We think that the likely range of advertising on cable, including local and classified advertisements on a multitude of local cable systems, will preclude the external pre-vetting of advertisements and we consider that a mechanism for dealing retrospectively with complaints of breaches of the code of practice will be sufficient. We deal with this in Chapter 7.

Pay-per-view

50. There is one other possible method of financing cable services to be considered. This is the question of premium payment for particular programmes (as distinct from channels) or, as it is usually called in North America, "pay-per-view". The technology of future cable systems, with the capability of all connected sets being individually addressable, will make it easy to direct the desired programme only to those who have paid for it or ordered it. But charging comparatively large amounts for individual programmes would have considerable social implications, particularly at a time when the majority of the country would not be cabled. For example, the BBC represented very strongly that one of their main concerns about cable competition lay in the possible siphoning of sporting events from free television to those who could afford to pay a premium for them and who happened to be in the area of a cable system. They were concerned generally that popular events might be bought up for subscription television, to the detriment of viewers who did not live in a cabled area or were either unwilling or unable to pay extra to watch what they had been accustomed to get as part of an ordinary broadcast service: but they identified as the critical part of this threat any introduction of pay-per-view on cable or DBS. The special danger of pay-per-view is that if there was a particularly popular event which enough people were prepared to pay a sizeable fee to watch—let us suppose that a total of one million people in a number of cable systems were prepared to pay £5 each—the revenue thus generated would be sufficient to guarantee that no ordinary broadcasting service, with an obligation to spend its available funds on continuous programming, could hope to compete for the television rights. In this way the ordinary viewer would be deprived of a major popular event in favour of a minority of the television audience. We found that this particular problem was already exercising many people we visited in the United States and in Canada and we have concluded that it would be safer for the time being to preclude pay-per-view programmes being offered on cable systems here.

51. Excluding pay-per-view will not necessarily prevent the siphoning of some events from broadcast television to subscription services, but it does at least avoid the problem in its most acute form. We do not think that pay-per-view is something on which cable expansion would rely heavily, and its exclusion will provide a measure of reassurance for public service broadcasting. The more general issue of cable programming and its potential effect on broadcasting, including the question of siphoning, is dealt with in the next chapter.

CHAPTER 6

CABLE PROGRAMME SERVICES

Number of channels

52. The role of cable systems in the United Kingdom so far has been clearly defined and very limited. They have been the means by which the three British broadcast television channels have been received by a small part of the population (currently just over 13 per cent of those holding television licences). Many television relay systems were built on the foundations of even earlier radio relay systems, some of which were established in the 1920s, and their common purpose was to bring television to people who could not at the time receive a satisfactory signal off-air or who, for local environmental reasons, were not allowed individual external aerials.

53. That limited role has not been through lack of aspiration to do more. Cable operators have had to work within a strict regulatory framework and with a licence from the Home Office which requires that they must carry all locally available public broadcasting services and may not—subject to very limited exceptions mentioned in the next paragraph—do anything else. Previous inquiries into broadcasting (like the Pilkington report in 1960 and the Annan report in 1977) have recommended that Governments should not allow cable operators to offer their own services to paying subscribers. Cable operators have therefore been forced to stick to their relay role and have consequently had no reason to install cable systems with capacity for many channels. Half the commercial cable systems in the United Kingdom can carry no more than 4 television channels, with 2 or 4 radio channels. Moreover, as constant improvements have been made in the broadcasting organisations' transmitter networks, cable's television relay role has become increasingly redundant.

54. The limited extensions of the local relay role have been of four kinds. First, the cable operator has normally been allowed, if he can receive off-air the signals from a neighbouring ITV region, to relay that channel as well as the local ITV service. Subscribers to some cable systems can therefore receive two or even three ITV channels as well as two BBC channels, and although ITV programmes may much of the time be identical, this does at other times increase the viewer's choice of programmes. Second, a limited pay-television experiment in London and Sheffield, based on a coin-in-a-slot meter, was authorised in the 1960s and ran between January 1966 and November 1968. Third, some community programming was authorised for cable television in 1972 and for cable radio in 1976. Only six community television stations and seven community radio stations have taken advantage of this authorisation and success has been difficult to achieve. One television station, at Greenwich, distributing three hours of community programmes a week, and five radio stations, at Basildon, Greenwich, Milton Keynes, Telford and Thamesmead, are all that remain operating. Fourth, thirteen pilot schemes of subscription television, the first starting in September 1981, enable a single channel to be used to provide an additional programme service, consisting almost entirely of cinema films.

55. So much for the past. We now have to consider the situation if cable operators are allowed to use modern technology to provide a large number of channels and programmes. The first point to make about the services on future cable systems is that we see no reason to restrict in any way the number of programme channels that can be supplied, subject only to the adequate provision of interactive channels of communication (see paragraph 5). Indeed, the object should be to provide the maximum variety and choice for the viewer. The questions we have considered relate more to the kind of programme services that cable will carry.

56. At this point we should perhaps say that we see no incompatibility between the expansion of cable services and the developing use of satellites for the distribution of television programme services. These developments are complementary to each other, particularly in the sense that cable systems could increase the number of viewers to whom DBS services are available and could enable them to be received more cheaply. Furthermore, under the 1977 Geneva Plan of the World Administrative Radio Conference, the United Kingdom has been allocated frequencies for only five national television channels by means of DBS and it could therefore give the viewer only a small part of the choice which wideband cable could offer. Moreover, DBS could not offer the interactive capabilities of modern cable systems.

57. A second kind of satellite, providing low-power telecommunications links rather than transmissions which can be received by the general public, is now widely used in the United States for feeding programme services to cable systems across the country. These telecommunications satellites are therefore an alternative not to local cable systems but to the trunk or microwave network which might otherwise provide the means of feeding the same programme services—particularly live programmes which could not be distributed in the form of videotapes—to a number of cable systems. The need to use satellites for this purpose is less in a country the size of the United Kingdom; moreover, we understand that there could well be problems within the United Kingdom in finding the necessary frequencies for an extensive satellite distribution network which would not interfere with existing and planned terrestrial services.

The "must-carry" rule for public service broadcasting

58. It is appropriate to start with the continuation of the relay role described above, which is usually referred to as the "must-carry" rule. All the evidence put to us assumed that a basic feature of a multi-channel cable system would be the carrying of public broadcasting services. There was no argument about whether or not this should be done, because cable operators, present and potential, considered that it would be in their interests to provide their subscribers with BBC and ITV programmes as part of their basic package covered by the fixed rental fee. The only question is whether there should be a formal obligation on them to do what they are likely to want to do anyway. We noted that this formal obligation exists in both Canada and the United States, despite the considerable freedom given to cable operators in other respects in the latter country. We think there are strong arguments for making it clear to the person who rents cable that in all circumstances he will continue to be able to receive the public broadcast services he has received in the past. We do not for example think that it should be open to an operator to remove a public service broadcast

channel provided as part of his basic package—say BBC 2 or Channel 4—in order to substitute a subscription service. Such action would be contrary to our basic philosophy that cable services are supplementary and not an alternative to public service broadcasting. We therefore recommend that cable operators should be required to carry all free broadcast television services serving their particular locality, whether present or future. Thus in addition to BBC1, BBC2, ITV and Channel 4 or the Welsh Fourth Channel, there should be a requirement to carry DBS services available to all viewers, but not those requiring payment of a subscription.

59. There is a temporary problem about the application of the must-carry rule to those existing systems which can carry only a few channels and which are no longer profitable on a relay basis only. Unless the operators of those systems can provide new revenue-generating services they may be forced to close down. On the other hand, if the operators can use channels for commercial purposes they can probably survive until modern cable systems can be installed. The Cable Television Association emphasised that it would not be in their members' commercial interests to seek to deprive their customers of BBC and ITV programmes. In many cases it would be possible to provide subscribers with the means of receiving broadcast signals off-air with television sets adapted for both off-air and cable reception, prior to utilising cable channels for other purposes. In the case of these out-of-date relay systems, we think it would be reasonable to give a temporary waiver of the must-carry rule provided the cable operator was willing and able first to provide the viewer at no extra cost to him with the means of receiving the public service channels satisfactorily off-air. The operator would require the consent of the franchising authority to the provision of new programme or other services on an existing system, and those services would then be brought within the oversight which we propose for new cable services. However, we think it important that this conditional waiver of the must-carry rule should not be used to perpetuate out-of-date cable systems and thus not achieve the other benefits of recabling with wideband systems. Accordingly, we recommend that this waiver should be for a period of no longer than five years. Nor could the existing relay company necessarily assume the automatic right to a franchise to install a modern multi-channel system in the area. It would obviously be very well-placed to get this but if there were other applicants they would have to be considered on a competitive basis.

60. We considered separately the application of the must-carry rule to radio services. Such a rule now exists, its form varying according to the capacity of the system. For example, in those systems which can carry only two radio channels in addition to television, the two services relayed must be chosen from the BBC's national radio services; if three can be carried, the operator may choose either another of the BBC's national services or a BBC or IBA local radio station serving that area; and if channel capacity is no problem, it is only after distributing all the BBC's national services and both local radio services, if there are two serving that area, that a cable operator can choose to relay any other authorised broadcasting station, such as Radio Luxembourg. It was suggested to us that radio relay by cable is nowadays of little practical significance because so much radio listening is from car radios and portable transistor sets. Nevertheless, we were told that both the Independent Local Radio contractors and the BBC attach importance to the carrying of radio by cable;

and, if we are moving towards a society in which many more activities in the home will use cable connections, it may well be that radio by cable will have a greater role than hitherto. We were not entirely persuaded by this argument but since the bandwidth for many radio channels is less than for one television channel and since on our visit to the Manhattan cable system it was noted that it carried under a must-carry rule 24 FM radio channels in addition to its 28 television channels, we suggest that the must-carry rule should continue to apply to new cable systems for radio.

Carrying of other United Kingdom broadcasting services

61. We turn from the broadcasting services which cable must carry to those which it may perhaps carry. We have already explained that some cable systems now carry out-of-area ITV services. This could develop considerably with interlinked multi-channel systems. However, the multiple relay of ITV services on a wide scale would raise sensitive issues for the ITV system, because the IBA awards contracts on the basis that the programme companies have the virtual monopoly (particularly of advertising) in their own area. This system could be distorted if, with the expansion of cable systems, a significant part of the population were able to choose whether to watch the local ITV service or that from another area. Nevertheless, we would support making it possible for the Welshman living in London to watch the Welsh Fourth Channel or the expatriate Scot to watch local services from Scottish Television, as part of the increased choice which cable will offer. We found a measure of support for this idea among some ITV companies although we accept that there could also be some problems. Moreover, given the fact that cable television is all about widening the viewer's choice we think it would be an anomalous arrangement if cable were able to offer its own programme services alongside the local ITV service but were not allowed to carry other United Kingdom broadcasting services. We think therefore that the relay of out-of-area ITV programmes is a matter which should be left for the cable operator to negotiate if he wishes to do so in the light of local circumstances and demand.

Carrying of foreign broadcasting services

62. There are also arguments for freedom to distribute foreign broadcasting services. Radio Telefis Eireann suggested to us that cable systems in the United Kingdom should be permitted to carry RTE's radio and television programming to increase the choice available for viewers and listeners, particularly those of Irish descent; and we noted that the proposals put forward by Merseyside Cablevision envisaged that RTE programmes would be included if they were allowed to install and operate a cable system in Merseyside. We see no objection to cable operators choosing to carry foreign services. It could be a help to international understanding and an aid to language education if some other European services were distributed here by cable, particularly when the development of DBS will make them more readily receivable in this country. We doubt however whether many cable operators will think the demand is likely to justify this, and again we think it should be left to them to decide.

63. In one significant respect, foreign broadcasting services are different from out-of-area ITV services. Since they originate outside this country, they are not susceptible to any British control. A case in point is advertising,

discussed in the previous chapter which concluded that advertising on cable should be subject to the same code of practice as is applied to Independent Television. We are aware that work is in progress towards European agreement on the standards to be applied to satellite broadcasting, as regards both advertising and programme material, and this may result in a measure of harmonisation. Nevertheless, and in the meantime particularly, it seems to us right that if a cable operator wishes to distribute a foreign broadcasting service he should not only ensure that he has the copyright owner's clearance but accept responsibility for its content.

Range and balance of programmes

64. The main potential threat to public service broadcasting from cable expansion is seen by the BBC and Independent Television to lie in what they regard as the commercial approach cable services are likely to take towards securing large audiences by popular programming, particularly by obtaining exclusive rights to programmes—popular sporting events especially—which will encourage people to pay for subscription services. The argument is that broadcasting will be both weakened by losing a large part of its audience and also impoverished by being deprived of certain programmes or by being forced to compete more on a popular level to the detriment of serious or quality programming. We do not ourselves rate this threat very highly but we must consider whether there should be any further safeguards, additional to the must-carry rule and the exclusion of pay-per-view, to minimise it.

65. One view put to us was that the value of cable expansion would lie in "narrowcasting", by which we mean its ability to provide television on a truly local scale and in such abundance that all kinds of specialist interests could find themselves catered for in a way that is not possible on broadcast television. It was sometimes implied that cable should confine itself—or be confined—to this role and not get involved in competing for mass audiences. Our assessment is that if left to develop freely, cable will not confine itself solely to a narrowcasting role. For example, there seems to be a clear and large market for the viewing of cinema films in the home. We find it difficult to characterise programming of this kind as of specialist interest or as falling within the description of "narrowcasting". The best of it will plainly appeal to a large audience which may be drawn away from some BBC and ITV programmes, just as we heard in the United States that a popular film on subscription television may draw a larger audience among those with access to it than any of the broadcast networks. However, only a small proportion of films will have this kind of drawing power and, assuming that the BBC and ITV continue to maintain the quality of their programmes, we do not think they need fear the loss of the bulk of their audience, even when cable eventually achieves a high penetration in this country. In any case, we doubt whether it is feasible to guard against this by preventing cable from trying to be popular. To do so would certainly inhibit the development of cable and, although we are aware that the Canadian authorities have decided that subscription television can go ahead only with restrictions on the proportion of box office successes which may be included among the cinema films shown, we came to the conclusion that it would be wrong to follow this precedent or to confine cable either to a narrowcasting role or to a heavily regulated regime. A sports or continuous news channel would similarly not come within the definition of narrowcasting.

Cable operators in our view should have freedom to provide programmes which are likely best to appeal to their customers, even though some of these may be provided by national programme providers to a large number of cable operators.

66. The converse to the argument that cable's role should be limited to certain types of programme is that it should be obliged to cover all fields, in the same way that public service broadcasting has an obligation to maintain a wide range and balance in its programming. The argument here is that public service broadcasting would be at an unfair disadvantage in competing with cable if it, but not cable, were to be required to provide a balanced service with, for example, a certain proportion of serious material and of adult education, some current affairs programmes or documentaries in peak viewing time, limitations on the number of films, quiz and game shows, etc. There could be no question of requiring a wide range and balance of programmes on individual cable channels and it seems to us artificial and unnecessary to place an obligation as to diversity on the totality of a cable system's output. It will happen anyway as a result of a multiplicity of channels and of the carrying of the public broadcasting services, and we do not believe that an attempt to spell this out in detail would represent any safeguard for the future of public service broadcasting. Indeed, we think it may be more realistic to think of the variety of services which could be provided by cable as in time reducing the need for some of the existing obligations on public service broadcasting, for example the amount of local programming by the ITV companies.

67. We therefore make no recommendation on the range and balance of programmes. Diversity is an aim that we think the franchising system can encourage. In any case, what is wanted and what can be financed may vary from one place to another, but there should be a presumption that all systems should provide children's programmes, education and a community service. We have already suggested that any cable franchise should be awarded after taking account of the various mixes of services to be offered. In this way, the franchise for that locality would go to the operator able to provide the best service, and we prefer this approach to one that seeks to lay down a lot of obligations on all cable operators from the start.

Exclusive rights

68. There is one particular threat to public service broadcasting, and thus to the viewer without cable, to be considered at this point. We have already referred (paragraphs 50–51) to the possibility that cable would have the money to negotiate exclusive rights for certain important sporting and national events. We were left in no doubt about the importance attached to this threat by the broadcasting organisations and also by a large number of our other witnesses since it would affect many viewers who cannot have or do not want cable. For this reason we have already recommended that pay-per-view should not for the time being be permitted as a source of cable finance. But is this a sufficient safeguard?

69. Section 30 of the Broadcasting Act 1981 enables the Home Secretary to make regulations to prevent the making of exclusive arrangements for the

broadcasting of sporting or other events of national interest. No such regulations have ever been made and the provision is regarded as a reserve power against the background of which the broadcasting organisations should agree among themselves about those events which should be protected against exclusive arrangements made by either the BBC or the ITV companies. The events which the BBC and the ITV companies have agreed should be protected in this way are the FA Cup Final, Wimbledon, Test Matches, the Derby, the Grand National, the Oxford and Cambridge Boat Race and the Commonwealth Games when held in the United Kingdom. However, it should be noted that despite the agreement in principle that exclusive arrangements for these events should be avoided, the position in practice is that most of them are covered by only one of the television networks, because the other does not think duplication is worthwhile.

70. Those who gave evidence to us from the cable industry disavowed any intention to seek exclusive arrangements for the kind of national sporting events that are already on the protected list. Nevertheless, we think it likely that cable will want to carry other major sporting events which will attract big audiences and thus encourage viewers to subscribe to cable services; and to draw a protective line around certain events could well have the effect of distorting the operation of the market and creating anomalies. However, the broadcasting organisations are likely to remain for the foreseeable future in a strong position to bid for the television rights to the majority of major events and we hope that one effect of the arrival of cable may be to extend the range of sports which the viewer will have the opportunity of watching and which will benefit from the injection of money from television rights. Admittedly, it could force up the fee for events which are already televised, but we had some sympathy with those who suggested to us that television at present secures sporting coverage on the cheap. Nevertheless, we think it right that the great national events should continue to be protected for the sake of the average viewer who has been used to seeing them on broadcast television. For this reason we take the view that the present list of protected events, modified from time to time if necessary and which may in the future need to be the subject of a formal regulation by the Home Secretary, should apply to new cable services as it applies to broadcasting. Apart from this, however, and subject to our proposed ban on pay-per-view services, we do not think it appropriate to place restrictions on the sporting or other events which cable television may wish to cover.

Local access

71. We have already stressed cable's important local role. We noted, for example, that in Canada an average of 8 per cent of cable revenue is spent on the production of community programmes. We have considered whether to recommend specific obligations on cable systems to provide facilities for local community programming and for access by local groups. This would allow local events and activities to be covered by television, enable local groups of various kinds to participate in the making of television programmes and provide a means of fostering interest in the community and encouraging expression within it. On balance we felt that this was another aspect of cable services that could be left to be taken into account in the franchising process, when the proposals of rival companies or consortia are considered. However, we believe

that there should be a presumption that the cable operator should accept responsibility for ensuring and financially assisting some community participation in cable programmes and we hope that cable systems' relationship with and contribution to their local communities might become a source of mutual pride.

Taste and decency

72. We now turn to the question of programme standards and in particular whether the traditional broadcasting requirements relating to taste and decency, and the suitability of programmes for children likely to be watching, should be imposed on cable operators. We think they should. In paragraph 11 of our report we explained why we did not accept the analogy with publishing, which is subject only to the laws on such matters as defamation, sedition and obscenity. The majority of those who gave evidence to us also took the view that the place of television in the average family home is such that restraints on pornography and violence are required additional to those imposed by the law of the land. We do not think that the mass of public opinion here would find acceptable some of the programmes receivable without restriction on cable channels in the United States. This may sound prudish but what is at issue is more a matter of the general climate of opinion than an argument about whether cable is more like publishing or broadcasting. In fairness to present and potential cable operators it should be said that many (but not all) of them agreed.

73. The question is however whether the industry should be left to operate a code of its own or whether a formal obligation should be placed on it. Frankly we doubt whether self-regulation would be acceptable on such a sensitive issue at a time when a large number of new cable operators, with different views on what is acceptable, will be establishing themselves on a local basis. Here again, it is clear that the majority of those who gave evidence to us shared this view.

74. Accordingly, and subject to the important exception dealt with in the next paragraph, we recommend that cable operators should be under the same formal obligations as the broadcasting authorities, namely to ensure so far as possible that nothing is included in their programmes which offends good taste and decency, is likely to encourage crime or lead to disorder or to be offensive to public feeling, and also to have special regard to programmes broadcast when large numbers of children and young persons are likely to be watching.

75. There is however a case for saying that there could be greater freedom for subscription channels which are optional extras to the basic package covered by the cable rental. The argument is that cable customers are not obliged to receive these channels and that those who wish to do so should be able to pay to receive material some of which might be offensive to others. In particular, relaxation of the times at which 'X' films can be shown would be of especial value to shift workers who would thus be able to watch during the day films that were otherwise available only in the late evening. The trouble about this argument is two-fold. First, subscription channels of films will carry much that would be offensive to no one and which all potential subscribers might wish to see. Second, we are concerned about the problem of confining so-called "adult" material to adults once a decision is taken to subscribe to a channel

which may carry it. However, we were impressed by the sophisticated decoder used by SelecTV in their pilot schemes of subscription television at Milton Keynes, Northampton and Tredegar. This enables each subscriber to lock out from his set films of censorship categories which are not wanted on the subscription channel for the time being, while allowing other categories to be received. Thus a parent going out for the evening can effectively prevent children watching an 'X' film or any film of either 'AA' or 'X' certificate or even, should it be desired, a film with an 'A' certificate. Any such system does of course depend on the parent bothering to activate the electronic lock and taking steps to maintain the secrecy of the personal code number, but otherwise the system is tamper-proof. We therefore take the view that on a subscription channel with sophisticated electronic locking facilities of the kind described, there would no longer be a need to restrict the time of day for showing the sort of films currently shown late in the evening because they are unsuitable for children. Furthermore, we think it would be acceptable with this safeguard to allow the showing, without any overriding constraints of taste and decency, of any film passed by the British Board of Film Censors for public exhibition with an existing censorship category[1]. We do not think it right to go any further and, for example, allow the showing of the kind of film which has hitherto not been passed for public exhibition and has been confined to cinema clubs or other video material which does not conform to public service broadcasting requirements of taste and decency.

Impartiality

76. Another question relating to programme standards is whether it is necessary on cable to require strict impartiality on matters of controversy. We are satisfied that news, whether it be national or local, should be presented with accuracy and impartiality. Comment, however, is a matter that we think could well be allowed greater latitude on cable, but not to the extent that there is a political bias across a cable system as a whole. Impartiality in community access channels should mean only impartiality in allowing access. Those allowed access would be under no obligation to be impartial. Similarly, individual channels could carry programmes provided by special interest groups, for example political or religious organisations. However, we think it would be wrong if the amount of programming by political parties and religious groups was solely dependent on their ability to raise money and for this reason we recommend that they should not have their own cable channels.

Foreign material

77. We have considered whether requirements should be imposed as to the proportion of British material which should be included on entertainment channels on cable. The broadcasting organisations are obliged to ensure that a "proper proportion" of their output is British (or from the European community generally) and they interpret this to mean that no more than 14 per cent should come from outside Europe. The subscription television pilot schemes were required in their cinema film programmes to show the same quota of

---

[1]"Existing censorship category" refers to the categories used so far by the British Board of Film Censors and any future equivalent categories, but not to the new "RESTRICTED (18)" category being introduced to cover some sexually explicit or violent films hitherto shown only in cinema clubs.

British films as applies to the cinema. However, we noted that the Government at the beginning of this year had reduced the Films Act quota from 30 per cent to 15 per cent and it has more recently announced that from 1 January 1983 the quota will be suspended altogether. The existence of the quota has not been effective in ensuring that sufficient British material has been produced to fill it, and this experience does not encourage us to propose that a quota should apply to cable programmes. Moreover, we received no evidence to suggest that a rapid expansion of the number of cable channels could in the short term rely very heavily on British material. Cable will inevitably have to buy in many programmes from elsewhere, mainly the United States, and unless it is able to do that, it will be much slower to develop. In the longer term, we have faith in the British film and television industry's ability to expand to satisfy the market provided the cable operators encourage this and do not continue to rely unduly on low cost material from abroad when there is sufficient British material available. The ITV programme companies are eager to have the opportunity to serve new outlets, as is the expanding independent production sector which has already been stimulated by the demands of Channel 4. We believe that the supervisory body might have a role in encouraging the production and use of British material on cable and possibly even have power to impose restrictions in the longer term, if that was thought appropriate, on the amount of foreign material which could be included in cable programme channels.

The cinema

78. We referred in paragraph 45 to the position of the cinema. A powerful plea was put to us by the cinema exhibitors for a measure of protection against the inroads of cable. We have already pointed out that one of the main attractions of cable lies in its provision of feature films, and it is clear that the development of the "cinema-in-the-home" concept may weaken the position of the cinemas, which have in any case been faced for many years with a steadily declining number of customers. What the cinema exhibitors would like in particular is a statutory restriction on the ability of cable operators to show new films, so that cinemas would retain the monopoly of exhibition for the first twelve months at least.

79. The normal practice at present is that cinema films are not shown on television for three years after release. This practice applies only because film producers and distributors have decided that it is in the best interests of the film industry generally that they should not make their films available to television during that time, to allow cinemas to exploit them fully. Exceptions are sometimes made. When the subscription television pilot schemes started, their wish to offer more recent films on payment was recognised but the cinema was still given some protection, through a condition of the cable operator's licence that a film should not be shown on cable within twelve months of its registration. The cinema exhibitors indicated to us that for cable they would be content with the twelve month rule where films were part of a subscription service but that otherwise they wanted the three year rule to apply, and they argued that these restraints should now be imposed by legislation rather than by agreement within the film industry.

80. We have some sympathy with the position of the exhibitors but we believe that market forces should be allowed to dictate the showing of films on

cable. Just as the showing of films on television now is determined by the film industry's assessment of its own interests, so we believe it should be left to decide on what terms it will make films available to cable. It may conclude that a film can be sold more profitably and established in the public mind by being premiered in the cinema; but if it ceases to believe in the commercial value of exhibiting films in the cinema before they are shown on cable television or on television generally, we are not convinced there is a case for imposing a statutory constraint on its freedom to sell its product in whichever market it chooses.

81. It was not our job to give detailed consideration to the problems of the cinema. If, however, our general approach is accepted we think that cinema owners have a point in arguing to us that they alone should not be forced to pay the Eady levy to help finance British film production. It is not for us to recommend whether or not the levy should be abolished or widened. But we think the arguments for equal treatment over this between the cinema and television—cable or otherwise—are made stronger by our conclusion that the cinema should not be accorded any statutory preferential position in relation to the availability of new films for exhibition.

Copyright

82. We received a number of submissions suggesting that a necessary precondition of the expansion of cable television should be the reform of the present laws on copyright. The Government is already considering reform of the copyright laws in the wake of the Whitford Report[1] and it was not for us to make proposals in that field. We have no doubt that the Government will give due weight to the extra need for amendment of the copyright law which may flow from a decision to allow greater freedom to programme services delivered by cable.

---

[1] Report of the Committee on Copyright and Designs Law, Cmnd 6732.

CHAPTER 7

OVERSIGHT

The nature of the task

83. In Chapter 2 we set out a broad perspective within which to see the future of cable television. To recapitulate briefly, it should be seen as supplementary, and not as a rival or alternative, to public service broadcasting. It should widen and enrich the viewer's choice by providing a large number of channels of special interest for which people are prepared to pay. It should be encouraged to be innovative, experimental and above all responsive to local needs which will vary from place to place. It cannot be run as though it were another branch of public service broadcasting. On the other hand its development will need oversight; and there must be some safeguards to preserve the quality and range of public service broadcasting, both as an end in itself and because many viewers, by choice or necessity, will remain solely dependent upon it.

84. The kind of oversight we envisage will involve both a positive and a reactive role. The positive aspects will be centred on the franchising process and on the choice of an operator to provide cable services in a particular area. Thereafter, the oversight will be reactive, intended to ensure that certain ground rules are observed but without constant supervision of the services that are provided.

The purpose of franchising

85. For the reasons given in paragraph 18, we feel strongly that there should be a formal process for granting franchises. A cable operator will have an effective local monopoly which should be conferred only after an opportunity for judging any competing bids and for securing the provision of the best service for the area concerned. Depending on decisions taken by the Government about the planning of the cable network, the franchising body might take the initiative in delineating areas and inviting applications for franchises; but given the local nature of cable systems we think there should also be room for cable operators to take the initiative in putting forward applications for franchises in areas of their choice. Any bids of this kind should however lead to the public advertisement of the franchise and the opportunity for other operators to put forward their own proposals to be considered in competition with the original application.

86. The decision on a franchise will involve the franchising body making a positive judgement about a number of elements in the applications before it and could also include some negotiation on them. These would include:

    (*a*) the area to be cabled. We are satisfied that there can be no precise answer as to the optimum size of a cable system, since it will depend partly on the nature of the area to be served. The ideal would be a franchise area of a size small enough to retain an identity with the locality and to enable cable to be installed expeditiously but which could at the same time support a wide range of cable services. It would be a pity if, by suggesting an arbitrary limit to the size of a franchise area, the effect was to deny to the suburbs of a city the benefit of connection to the cable system serving the city itself. However, because we attach importance to the local nature of cable systems, we take the view that even in the conurbations a franchise area should not cover more than about half-a-million homes, and should normally be smaller than that. Equally so, if cable is to spread successfully, many relatively small towns should be cabled even though they may not be able to support such a large number of channels. There are two other considerations relating to the area to be cabled. The franchising body will need to take into account the comprehensiveness of the cabling within the area concerned, so that it is not restricted to streets where a majority of occupiers are likely to pay for cable. It could also speed up the widespread development of cable and avoid what is known as "cherry-picking" by seeking in some cases to combine less attractive localities with those carrying the best commercial prospects or even by asking large consortia which bid for a prosperous area to provide a separate cable system in a less promising area elsewhere;

(b) the speed with which the cable system is to be installed and, to learn a lesson from United States experience, the sequence in which different parts of it are to be cabled, in order to avoid the situation in which the cable system is never in practice extended beyond the most profitable parts of the franchise area;

(c) the ownership of, and the interests represented in, the prospective cable operating company and any problems of monopoly whether national or local;

(d) the channel capacity of the system and the range and diversity of the proposed programme channels, including the proposed financing arrangements;

(e) the arrangements proposed for community programmes and local access;

(f) the intentions regarding the provision of interactive services.

87. We considered what the term of the franchise should be. Because much of the evidence given to us has argued for the need for a lengthy period of assured use over which the very considerable cost of installing the cable can be recouped, we in turn stress that we do not propose the licensing of the cable provider and that the franchise we are considering is solely for the operation of the system. If the cable operator's franchise came to an end, this would not render the system itself useless and of no value: it would be available for use by another operator. Indeed, we made clear in paragraph 22 that where the operator also owned the cable he would be required in the event of his losing his franchise to sell or lease his infrastructure on a predetermined basis to another operator in order to ensure continuity of service to the cable customer. In these circumstances we consider that the cable operator should ordinarily be granted a franchise, as are ITV programme contractors, for eight years. However, in the first instance, when the cable system and its services have to be developed from scratch, it should be for ten years from the start of operation.

88. Existing cable operators should be allowed to continue operating until a full franchise is granted for the area covered by their system, whether to them or to another operator; and of course the grant of a franchise would not affect the continuation in the same area of non-commercial master antenna systems providing nothing but broadcast relay (such as those which simply use one aerial to feed BBC and ITV programmes to all the occupants of a block of flats).

Subsequent oversight

89. Once the franchising body had taken its decision on the best cable operator for a particular area, its most positive function would be completed. It should remain in the background unless important developments or specific complaints required it to take an interest. We see no need for it to control the charges to cable customers. Nor would there be arrangements for continuing regulation by way of approval of programme schedules, pre-vetting of advertisements or constant scrutiny of output. Cable systems should operate under minimum constraints. The approval of the franchising body would however be needed if it was subsequently proposed to modify the basis on which the franchise had been granted, for example if there was a change of ownership of the cable operating company or if there were significant modifications of programme channels, as we recognise might happen in the light of experience and the availability of new services. We have already made clear the extent to which we think that cable programmes should conform to the traditional requirements as to taste and decency, and also that advertising should conform to the IBA code of advertising standards and practice. There is therefore a need for an authority which would keep in touch with what is going on; serve as a forum of advice to operators; and be generally known as the body to receive and adjudicate on complaints concerning the service being provided if the viewer felt the operator had given an inadequate response. We have also identified the need for an eye to be kept on the amount of United Kingdom programme material being used and on the way in which advertising on cable might develop. Above all, however, it should be responsible for judging whether cable operators were living up to their promises and for responding in a flexible way as the cable industry developed.

Local or central franchising?

90. There is therefore a crucial link between oversight and franchising, and the first question is who is to do the latter. Despite the vigorous case put forward by the Association of District Councils in particular, the bulk of the evidence given to us was against franchising being undertaken by local authorities. We agree with this view. Cable systems may cross local authority boundaries and not be readily susceptible to control by one authority alone; it might be desirable to combine for franchising purposes less attractive areas with those with the best commercial prospects; and our examination of local authority franchising in the United States does not lead us to think it is a good model for this country. In particular, we noted the growing feeling in the United States that local authorities have been using their powers (which are based on no more than the same need to obtain wayleaves to dig up public highways as exists in this country) to excess, slowing down the franchise process and demanding too much from cable operators. Furthermore, local franchising

would not be at all an appropriate way by which to facilitate the speedy development of cable systems throughout the country in a way which allows some overall oversight of what is happening.

91. We therefore recommend that franchising should be undertaken centrally. Central franchising has the advantage that it greatly facilitates the link we wish to see between franchising and oversight, because our view is that the latter can only be undertaken nationally and not in isolation in each individual area. Indeed, if the local authorities were to be responsible for franchising we think there would also have to be some national oversight body, thus involving some quite unnecessary duplication of functions. Central franchising therefore enables promise and performance to be judged by the same body. It also greatly simplifies the question of dealing with unsatisfactory performance because the ultimate sanction is to take away a cable operator's franchise. We discuss sanctions further in paragraphs 99–101.

92. If franchising is to be a central function, we are clear that the franchising body should nevertheless devise ways to obtain local opinion on rival bids, as is currently the practice in relation to ITV and ILR. Some means may also have to be found to prevent the grant of local authority wayleaves being used as an indirect means of controlling cable systems. We understand that it has been common for local authorities in granting wayleaves to impose conditions requiring the cable operator to pay continuing fees based on a percentage of the system's revenue and to obtain their approval to any increase in subscriber charges, whereas we see no need to control charges. Once cable is allowed greater freedom the scope for further conditions to be imposed would be increased. We therefore take the view that either the central franchising body should have power to grant wayleaves or, if local authorities are to retain the power, the franchising body should have reserve powers to override the local authority if need be.

The cable authority

93. In our view there are only two candidates for a central franchising and oversight body—the IBA with different requirements placed on it in respect of cable television, or a new body.

94. It would be possible for the IBA, which is at one remove from the programme companies and already also looks after ILR as well as ITV, to oversee different regimes for broadcasting and cable. It has great experience in handling franchise applications, administering an advertising code and setting technical standards. It also has a network of regional offices and advisory committees. There is moreover a case for saying that the very fact that it exists would mean that it could take responsibility for cable, under different ground rules from broadcasting, more quickly and smoothly than a new body which would have to start from scratch.

95. Nevertheless, we do not think this would be the best solution. However conscientiously the IBA approached its differing responsibilities, there would be at least an appearance of some conflict of interest which could become real if, contrary to our expectations, cable made serious inroads into ITV or ILR

advertising revenue. In any event there would be suspicion that the IBA might have an over-protective attitude to public service broadcasting, and this could deter potential investors in cable. We have expressed the hope that cable will most effectively widen viewers' choice by being innovative and developing in a distinctive way free of the kind of perceptions which presently govern broadcasting and the press; this can best be achieved if its supervision is independent of those who supervise the existing media. We think that the additional work load for the IBA would also be considerable; and it has no expertise in the information technology aspects of cable which should benefit business and the consumer. The evidence we received was overwhelmingly against the IBA being responsible for cable.

96. We therefore recommend the establishment of a new cable authority which should be responsible both for awarding franchises and for monitoring performance.

97. We envisage its governing body having membership based on wide-ranging interests rather that narrow interests relating solely to cable and other media. We see value in ensuring that the interests of the different parts of the United Kingdom are represented so that the desirability of spreading cable throughout the country—and its effects on those areas without it—can be understood from the start. The authority will need a small staff sufficient only to help it with the franchising process and in those aspects of oversight which we identify as essential. Its costs should be modest but will need to be met by appropriate charges.

98. The new authority will need to be established by legislation: and we do not know whether time for this can be found in the next session of Parliament. If legislation in the 1982/83 session is not practicable it would be possible, following discussion and debate of our report, for the Government temporarily to use the powers described in paragraph 18 to issue licences in order to facilitate an early start on the new cable systems. If however this course is adopted, we hope it would be accompanied by the establishment of the new cable authority initially on a non-statutory basis so that it can play its part from the outset and on the understanding that, unless there were exceptional circumstances, Ministers would use their licensing powers in accordance with its advice.

Sanctions

99. We have expressed our conviction that cable should develop with only the minimum of constraints; but we have made it clear that cable operators should be subject to certain obligations. Their programmes will have to comply with the ground rules we proposed in Chapter 6 and their advertisements will have to conform to a code of practice. They will also be expected to live up to the offers made in bidding for the franchise. The cable authority will have oversight only from a distance, but it will consider complaints and their remedy. How in the last resort should its supervision be enforced?

100. We have referred in paragraph 91 to deprivation of franchise being the ultimate sanction. The existence of this power will in most cases be sufficent. We think it would be rare for the power to have to be exercised, but we have to

consider whether this knowledge would weaken the power of the cable authority vis-à-vis an operator who, for example, was committing persistent but relatively minor breaches of the taste and decency requirements or who was making no real attempt to show an adequate amount of British programme material. Do we need the kind of intermediate sanction which exists in other countries to deal with repeated complaints of misbehaviour that could scarcely justify the franchise being revoked? Even if the complaint is not serious enough to justify the franchise being revoked, fear that persistent breaches will reduce the chance of its being renewed is likely to encourage most operators to co-operate. But there may be cases where this is not enough: the franchise may have some years to run and thoughts of renewal may be remote; or the operator may either not wish to continue the franchise or have already given up hope of its being renewed.

101. We have considered the possibility of the cable authority being authorised to impose financial penalties on the operator which could, as in the United States, be taken from a performance bond lodged by him, but we do not think this is a very satisfactory solution: performance bonds in particular are not well suited to failings which are a matter of qualitative judgement, as breaches of taste and decency would be. An arrangement we favour is one in which the operator would forfeit his deregulated position until the cable authority was satisfied that he could be relied upon to meet his obligations without supervision. Thus in the light of persistent complaints and failure by a cable operator to remedy their cause, the cable authority would be empowered to issue a public warning that recurrence would lead to the operator being brought under closer supervision. For example, he could be required to submit his programme schedules or the details of particular programmes or his advertisements before they were shown. The cable authority would have discretion to apply this across the output of the cable system or restrict it to particlar channels; and it would have power to direct that certain programmes should not be shown or even that particular channels should be closed down. The public nature of the advance warning that was given would act as a restraint on unnecessarily interventionist action by the cable authority; but from the cable operator's point of view the imposition of a regulated regime for a period would be a considerable inconvenience and in our view the kind of sanction which, coupled with the threat of revocation of a franchise, would be sufficient.

CHAPTER 8

SUMMARY OF CONCLUSIONS AND RECOMMENDATIONS

102. Our terms of reference told us to assume the Government's willingness to consider an expansion of cable systems for entertainment and other services, but in a way which would safeguard public service broadcasting; and asked us to make recommendations on the questions affecting broadcasting policy which would arise, including in particular the supervisory framework. We were not therefore concerned with the cable technology to be used or with the provision of services on cable which would have no impact on broadcasting policy. Our task was to consider whether arrangements can now be devised under which cable television and public service broadcasting can co-exist without unnecessary inhibitions on the development of the former and without damage to the essentials of the latter. We believe that they can within the broad perspective we have described. We have considered the potential of cable and the circumstances in which it is most likely to develop. Equally, we have considered the social implications of a situation where some people—either because of their means or because, for a long time at any rate, of where they live—would be able to pay for services which were not available to others. We see no convincing argument against this provided the result is not a deterioration in the range and quality of the broadcasting services freely available to all. We think such a result can be avoided; but to provide both positive encouragement to new multi-channel cable operators and at the same time safeguards for public service broadcasting and those who rely on it involves striking a nice balance. We therefore emphasise the importance of looking at our main conclusions as a whole.

103. Against that background we draw together here our conclusions and the recommendations we have made in the course of our report.

Functions in cable and the case for a franchise system
1. We identify four functions in cable (paragraph 15):
    (a) *the cable provider*—the installer and owner of the physical infrastructure;
    (b) *the cable operator*—the manager of a local cable system who puts together a package of cable services to sell to customers, from whom he collects revenue;
    (c) *the programme or service provider*—the assembler of programmes into channels or segments of channels for sale to cable operators, or the provider of other services;
    (d) *the programme maker*.

2. The key figure of these four will be the *cable operator*. The initiative for the installation of cable system is most likely to come from him. In any case, he will be answerable to his customers for the service he provides and will be the key figure for the purposes of any supervisory framework (paragraphs 16–17).

3. Because cable systems will have a *de facto* monopoly, there should be a formal franchising system in which franchises for cable operators should be open to competition (paragraph 18).

4. It is not necessary for the *programme provider* or *programme maker* to be licensed; the cable operator should be held responsible for the programmes and services he distributes, other than those for which a United Kingdom broadcasting authority takes responsibility (paragraph 19).

5. A separate licence for the *cable provider* might not be necessary if the onus is placed on the cable operator to ensure that any system he operates meets the appropriate technical specifications (paragraph 20).

6. There is no need for separation of ownership between cable operator and cable provider; but where the cable operator is also the cable provider he should be required if deprived of his operator's franchise to sell or lease his infrastructure on a predetermined basis to another operator (paragraph 22).

7. The cable operator should be free to provide programmes if he wishes; but care should be taken in awarding franchises to encourage a diversity of programme sources and avoid any undesirable monopoly. There should be an expectation that some channels should be available for leased use (paragraph 23).

Ownership of cable operators

8. Central and local government, political parties and organisations, and religious bodies should be excluded from direct participation in any form in the ownership of companies operating cable systems (paragraph 25).

9. The press, Independent Television and Local Radio contractors and foreign companies should be allowed to participate in the ownership of cable operating companies, but individual companies of this kind should not be permitted to hold a controlling interest (paragraph 26).

10. It is desirable but not essential that there should be some local participation in the ownership of cable operating companies (paragraph 27).

11. The franchising authority would be able, in the light of local circumstances, to consider whether a particular level of ownership or involvement would have undesirable monopoly implications (paragraph 28).

12. There is no reason why a particular company should not hold more than one franchise. There is little likelihood of a national monopoly arising but the franchising authority could monitor the situation (paragraph 29).

13. There should be no restrictions on the ownership of programme providers but (see paragraph 76) political parties and religious groups should not have their own channels. There should be no restrictions on the ownership of cable providers (paragraph 30).

Cable's sources of income

14. The first source of finance should be the rental charge for a basic package of cable services (paragraph 32).

15. Subscription for particular additional channels should be allowed (paragraph 35).

16. Substantial cable expansion in the United Kingdom is unlikely to be able to take place solely on the basis of rental and subscription (paragraph 35).

17. Because of its ability to provide new opportunities to both local advertisers and national advertisers seeking specialised or local audiences, cable could encourage new spending on advertising and should benefit the consumer (paragraphs 42–43).

18. To enable cable to achieve its full potential, advertising should be permitted (paragraph 47).

19. It is unnecessary in the early years of cable's development to limit the amount of advertising permitted. This would be unlikely to damage ITV or ILR to any significant extent in the short or medium term. There should be flexibility through the franchising process to impose restrictions at a later date if this proves necessary (paragraphs 47–48).

20. Advertising on cable should conform to the same standards and code of practice as apply to ITV; but there is scope for the sponsorship of programmes on cable provided that certain rules including the separation of advertisements from editorial matter are observed. The external pre-vetting of advertisements will be impracticable and a mechanism for dealing retrospectively with complaints will be sufficient (paragraph 49).

21. Pay-per-view should not be permitted for the time being as a method of payment for individual cable programmes (paragraph 50).

Cable programme services

22. Cable operators should be allowed to provide however many new programme channels they choose (paragraph 55).

23. We see no incompatibility between cable expansion and the development of direct broadcasting by satellite (paragraph 56).

24. New multi-channel cable systems should be required to carry all existing and future free BBC and Independent Television and Radio services serving the area concerned (paragraphs 58 and 60).

25. Existing limited-capacity cable systems could be excused this requirement for a period of no more than five years on condition that the operator can and does provide his customers with the means to receive satisfactory signals off-air at no extra cost to them. With the consent of the franchising authority, an existing cable operator could then provide new programme and other

services on his system pending the award of a new franchise for the area concerned. He would not however have an automatic right to a new franchise for the areas covered by his cable system (paragraph 59).

26. Cable operators should be free to negotiate the relay of any out-of-area ITV services (paragraph 61).

27. Cable should be free to relay foreign broadcasting services with the agreement of the copyright owners (paragraph 63).

28. Cable programme services should not be confined to a "narrowcasting" role. Conversely they should not be subject to the public service broadcasting requirements regarding range and balance. The diversity of services to be carried should be taken into account in the franchising process (paragraphs 65–67).

29. Cable should not be allowed to obtain exclusive rights for national sporting events of the kind which are protected under section 30 of the Broadcasting Act 1981. These events may need to be the subject of a regulation by the Home Secretary and the list may need to be revised from time to time (paragraph 70).

30. There should be no specific obligations on cable systems to provide facilities for community programming and local access but there should be a presumption that these would be provided and proposals for such facilities should be taken into account in the franchising process (paragraph 71).

31. For the reasons given in paragraph 11, we do not think that at the present time cable television can be regarded as just another branch of publishing, subject only to the law of the land, or that self-regulation would be acceptable. Cable programmes, with the single exception covered in the next recommendation, should be subject to the same obligations as the BBC and IBA not to offend good taste or decency, to be likely to encourage crime or lead to disorder or to be offensive to public feeling, as well as to have special regard to programmes broadcast when large numbers of children and young persons are likely to be watching (paragraphs 72–74).

32. However, on any channel for which a specific subscription is paid and which is capable of being electronically locked by the subscriber there should be no restriction on the time for showing the sort of films currently shown late in the evening because they are unsuitable for children. Additionally, we think it would be acceptable on such channels to show at any time any film passed for public exhibition by the British Board of Film Censors in an existing censorship category (see footnote to paragraph 75). But we would not go further in allowing the showing of other films or video material which do not conform to traditional taste and decency requirements (paragraph 75).

33. News should always be presented impartially. Otherwise, impartiality on individual channels need not be required but a cable system as a whole should not maintain a position of bias. There should be impartiality of access to local cable channels (paragraph 76).

34. There should for the time being be no minimum quota of British material but the use of British material should be encouraged (paragraph 77).

35. There should be no formal restrictions for the sake of the cinema industry on the films which may be shown on cable, but there is a case for equality of treatment as to the requirement to contribute to the British Film Fund (paragraph 80–81).

36. The Government should consider the implications for the existing copyright law of a decision to allow greater freedom to programme services on cable (paragraph 82).

Oversight

37. Oversight has both a positive and a reactive role. The positive aspects should be centred on the franchising process. Thereafter, oversight should be reactive, intended to ensure that certain ground rules are observed but without constant supervision of the services provided (paragraph 84).

38. Although the franchising body may take the initiative in delineating franchise areas and inviting applications, it should also be possible for cable operators to take the initiative in applying for franchises in areas of their choice (paragraph 85).

39. The franchise decision should take account of (paragraph 86):
 (*a*) the size of the area, for which we have suggested certain criteria but recommend no arbitrary limits; the comprehensiveness with which all streets within the area will be cabled; and in some cases the desirability of encouraging the cabling of less commercially attractive areas along with those with the best prospects;
 (*b*) the speed of installation of the system and the sequence in which different parts of it are to be cabled;
 (*c*) the ownership of and interests represented in the prospective cable operating companies and any problems of monopoly;
 (*d*) the channel capacity and the range and diversity of programme channels, including the financing arrangements;
 (*e*) arrangements for community programmes and local access;
 (*f*) the intentions regarding the provision of interactive services.

40. The term of the franchise should ordinarily be eight years but should in the first instance be ten years (paragraph 87).

41. Existing cable operators should be able to continue pending the grant of a franchise for their area. Master antenna cable systems providing only broadcasting relay would be able to continue notwithstanding the grant of a multi-channel cable franchise for the area concerned (paragraph 88).

42. Once a franchise has been awarded oversight should be on a monitoring and reactive basis only. There is no need to control charges to cable customers. Programme schedules should not require advance approval. Advertisements

should not be subject to external pre-vetting. There should be no constant scrutiny of output (paragraph 89).

43. The approval of the franchising body should be required for any subsequent change in ownership of a company operating a cable system or significant modifications in programme channels (paragraph 89).

44. There is a need for a body to keep in touch with what is going on; serve as a forum of advice to operators; be generally known as the body to consider complaints; keep an eye on the use of United Kingdom programme material and developments in advertising; judge whether cable operators are living up to their promises; and respond flexibly as cable develops. This would have to be done by a central body (paragraphs 89 and 91).

45. Franchising should also be undertaken centrally rather than by local authorities. Promise and performance can then be judged by the same body. The central franchising body should however devise ways to obtain local opinion on rival bids. The franchising body should have power to grant wayleaves or it should have reserve power to override the wayleave powers exercised by local authorities (paragraphs 90–92).

46. A new cable authority should be established and given responsibility for awarding franchises and monitoring performance. Its governing body should have a membership based on wide-ranging interests rather than one relating solely to cable and other media, and the interests of different parts of the United Kingdom should be represented. The cable authority should be a statutory body. If necessary it should be established initially on a non-statutory advisory basis (paragraphs 96–98).

47. The ultimate penalty for any cable operator who was grossly in breach of his obligations would be loss of his franchise (paragraph 100).

48. A less drastic sanction will be needed for use in circumstances not warranting loss of the franchise. Financial penalties are not recommended. A better course would be to give power to the cable authority to impose a regulatory regime on a cable operator for a period by requiring the submission of programme schedules or the advance vetting of programmes or advertisements (paragraph 101).

APPENDIX A

CONSULTATION DOCUMENT ISSUED BY THE INQUIRY ON 7 APRIL 1982

INQUIRY INTO CABLE EXPANSION AND BROADCASTING POLICY

CONSULTATION ABOUT THE ISSUES UNDER INQUIRY

The Inquiry has been asked by the Home Secretary

"To take as its frame of reference the Government's wish to secure the benefits for the United Kingdom which cable technology can offer and its willingness to consider an expansion of cable systems which would permit cable to carry a wider range of entertainment and other services (including when available services of direct broadcasting by satellite), but in a way consistent with the wider public interest, in particular the safeguarding of public service broadcasting; to consider the questions affecting broadcasting policy which would arise from such an expansion, including in particular the supervisory framework; and to make recommendations by 30 September 1982".

The background is that so far governments in the United Kingdom have adopted the policy that broadcasting should be conducted only as a public service by public authorities set up for the purpose. The only authorities licensed to broadcast (the BBC and the IBA) are placed under an obligation to provide services for the dissemination of information, education and entertainment which maintain a high general standard both technically and in their content, a proper balance and a wide range in their subject matter. They also have specific obligations in relation to programme standards: to ensure so far as possible that nothing is included in their programmes which offends good taste and decency, is likely to encourage crime or lead to disorder, or to be offensive to public feeling; and that due impartiality is preserved in the presentation of news and in the treatment of controversial matters.

Hitherto the justification for this kind of regulation has rested on two features of broadcasting:
- (a) it involves the use of a limited resource—transmitting frequencies—whose allocation is the subject of international negotiation between governments;
- (b) it is a powerful medium which is brought direct into people's homes, with great potential to influence or offend them there.

The part played by cable systems to date has fitted into this framework in that the essential function of the systems has been to relay BBC and ITV (and independent local radio) programmes. The extension of the provision of other programme services by cable systems would bypass the shortage of radio frequencies and the main need for regulation described above would not therefore arise. Furthermore, if cable systems are to be developed for additional services, there are economic and other arguments for giving the maximum incentive by providing a wide choice of programmes as quickly as possible. Indeed, one view is that broadcasting by cable should be regarded simply as another branch of publishing, with no more control or restriction than applies to the written press—for example, the restrictions on such matters as defamation, sedition and obscenity which are imposed by law. If this view is accepted, there is no need for regulation at all, though a voluntary code, possibly monitored by a body like the Press Council, would not be ruled out.

On the other hand it can be argued that the expansion of cable services would not remove all need for some degree of regulation. Broadly these arguments are that
- (a) cable is different from the written press, not only because there is almost bound to be an effective monopoly in any given local area, but also because a large

part of the country will continue to depend on off-air services for the foreseeable future;

(b) an expansion of programme services by cable could damage the quality and range of public service broadcasting, on which viewers who cannot receive or do not wish to pay for cable services would continue to depend, by obtaining exclusive rights to national and sporting events etc or by attracting audiences (and in the case of ITV advertising revenue) away from broadcast services and leaving them less able to provide the range of programmes now offered.

(c) regardless of the medium of transmission, television programmes brought direct into the home have a power, an intimacy and an influence which justifies supervision.

These questions relating to whether there is a need for regulation are central to the Inquiry's task, and it would welcome views and comments on the way in which the question of regulation should be approached and on the supervisory framework (if any) which might be appropriate.

Within this general question, a number of more particular issues arise on which the Inquiry would like to have views. Among these are:

1. Should there be restrictions on the scale or ownership of cable companies, eg to prevent excessive monopoly or to exclude or limit foreign interests, political or religious groups, the press or existing broadcasters?

2. Should there be separation between the cable operator and the programme provider, in order to facilitate a multiplicity of programme providers on each cable system? This arrangement would also make it easier, in cases of unsatisfactory performance, to exclude a programme provider without robbing the cable customer of all home entertainment services. Or is the incentive to cable and the demand for connection so closely related to the programmes to be provided that the cable operator cannot realistically be asked to forgo the opportunity to provide programmes?

3. What should be the basis of finance of cable entertainment? Should it be subscription, or should cable companies be permitted to finance themselves wholly or partly from advertising? If advertising were to be allowed should it be subject to control and supervision (cf the IBA system)? What repercussions would follow for independent television, independent local radio or local newspapers?

4. Should the present obligation on cable operators to relay the broadcast services of the BBC and IBA be retained? Or is there a case for removing it altogether or modifying it or waiving it for a limited period (the last course being intended to allow cable operators to maximise revenue from their existing limited capacity cable until they can lay down wide-band cable which will have adequate capacity for broadcast services along with others)?

5. Should cable operators be under an obligation (as the broadcasting authorities are) to provide a wide range and balance of programmes, including programmes for minority interests? If so, would it be reasonable for the obligation to apply to the totality of programme channels on a cable system rather than to each channel individually?

6. Should the traditional broadcasting requirements relating to taste and decency and the suitability of programmes for children likely to be watching be imposed on cable services? If certain channels can be received only by payment of an additional fee should the operator be free to supply on those channels programmes which might offend others, subject only to the application of the criminal law? Is there scope for specific rules related to the censorship category

of cinema films (as in the subscription television pilot schemes, which preclude 'X' films being shown before 10 p.m.)?

7. Should the traditional broadcasting requirements relating to impartiality be imposed on cable services, or should it be possible for particular interest groups to provide programmes or even to run individual channels?

8. Should there be formal safeguards against cable systems obtaining exclusive rights to certain events (along the lines of those in section 30 of the Broadcasting Act 1981 restricting exclusivity by one of the broadcasting authorities) or otherwise intended to prevent the impoverishment of broadcast services?

9. Should there be any restriction on the relay of foreign broadcasting services or on the proportion of foreign programme material included in cable services?

10. Should there be any protection for the cinema industry through a restriction on the showing on cable television of new films? (Films are not normally shown on television until they are three years old; and the licence for the subscription television pilot schemes precludes the showing of films within twelve months of their being first registered.)

11. If there were to be a need for a supervisory framework other than self-regulation, what should be the means of supervision? Is there scope for local supervision, or should it be carried out by some national body? Could the responsibility appropriately be given to an existing body or is a new public body needed?

12. What powers and sanctions would be necessary or desirable to ensure compliance with any scheme of regulation which might be adopted?

These questions are not intended to limit the scope of the representations which the Inquiry is inviting, but merely to help to direct comments towards the issues which the Inquiry thinks it will need to resolve. General views and constructive proposals on these and any matters within the Inquiry's terms of reference should be sent to the Secretary to the Inquiry at the address below as soon as possible. It would be helpful to the Inquiry to know in the meantime which of the bodies receiving this invitation proposes to submit evidence, which should be received preferably by 31 May 1982 at the latest.

J C Davey
Inquiry into Cable Expansion and Broadcasting Policy
Whittington House
19–30 Alfred Place
London WC1 7EJ

APPENDIX B

LIST OF ORGANISATIONS AND INDIVIDUALS WHO MADE WRITTEN SUBMISSIONS OR WITH WHOM DISCUSSIONS WERE HELD

ABC Video Enterprises
Advertising Association
Advertising Standards Authority
Age Concern
Air Call Communications
An Comann Gaidhealach
Arts Council of Great Britain
Association of Broadcasting Staff
Association of Cinematograph, Television and Allied Technicians
Association of District Councils
Association for Film and Television in the Celtic Countries
Association of Independent Radio Contractors
Association of Metropolitan Authorities
Association of Scientific, Technical and Managerial Staffs
Dr Robert Batscha, Museum of Broadcasting, New York
Beecham Products
Birmingham Post and Mail
Professor John Bowker
Dr J O Boyd-Barrett
British Academy of Film and Television Arts
British Actors Equity Association
British Aerospace
BBC
British Board of Film Censors
British Copyright Council
British Council of Churches, Division of Community Affairs
British Film Institute
British Film and Television Producers Association
British Phonographic Industry
British Radio and Electronic Equipment Manufacturers' Association
British Telecom
British Videogram Association
Broadcasting Research Centre
Broadcasting Research Unit Working Party on New Technologies
Mr Les Brown, Editor, "Channels" Magazine, USA
Cable Television Association of Great Britain
Cablecasting
Canadian Association of Broadcasters
Canadian Broadcasting Corporation
Canadian Cable Television Association
Canadian Radio—Television and Telecommunications Commission
CBS
Central Religious Advisory Committee
Centre for Mass Communication Research
Centre for Policy Studies
Channel 4 Television Company
Charterhouse Japhet
Church of England
Cinematograph Exhibitors' Association
Cinematograph Films Council
CJOH-TV, Ottawa

Mr R Collins
Comcom
Co-Media Video Publications
Comet Group
Confederation of Aerial Industries
Confederation of British Industry
Confederation of Entertainment Unions
Conference of Socialist Economists, Communications Working Group
Consumers' Association
Council for Educational Technology
Sir Geoffrey Cox
Mr Tim Clement-Jones
CTVC
D'Arcy—MacManus and Masius
Deaf Broadcasting Campaign
Deloitte, Haskins and Sells
Department of Communications, Canada
Department of Education and Science
Department of Education and Science (Office of Arts and Libraries)
Department of Industry
Department of Trade
Dorland Advertising
Educational Television Association
Professor Andrew Ehrenberg and Mr T P Barwise
Mr Harold Evans, Goldcrest Films
Electrical, Electronic, Telecommunication and Plumbing Union
Eosys
Federal Communications Commission, USA
Free Church Federal Council
Professor N R Garnham
Dr Henry Geller
General Electric Company
Mr Kevin Goldstein-Jackson
Sir Philip Goodhart MP
Granada Group
Professor Nigel Grant
Greater London Council
Greenwich Cablevision
Mr Eldon Griffiths MP
Guild of British Newspaper Editors
Mr William A Henry III, "Time" Magazine
Home Box Office
Home Office
Incorporated Society of British Advertisers
Independent Broadcasting Authority
Independent Programme Producers Association
Independent Television Companies Association
Independent Television News
Institute of Information Scientists
Institute of Journalists
Institute of Practitioners in Advertising
International Thomson Organisation
"It is Written"
Mr Michael Johnson
Justice
Ladbroke Group

Lancashire Community Cable
Mr Colin W Lecky Thompson
Liberal Party
Link House Communications
London Hydraulic Power Company
London Transport
Mail Order Traders' Association
Manhattan Cable Television
Mr Roger Marshall
Mechanical Rights Society
Mercia Sound
Mercury Communications
Merseyside Cablevision
Mersey Television Company
Methodist Church, Division of Social Responsibility
Mirror Group Newspapers
Musicians' Union
National Association of Broadcasters, USA
National Association of Conservative Graduates
National Board of Catholic Women
National Cable Television Association, USA
National Computing Centre
National Consumer Council
National Economic Development Office, Electronics Consumer Goods Sector Working Party
National Economic Development Office, Information Technology Sector Working Party User Panel
National Federation of Women's Institutes
National Union of Conservative and Unionist Associations, Women's National Advisory Committee
National Union of Journalists
National Viewers' and Listeners' Association
Nationwide Festival of Light
NERA International
New York City Bureau of Franchises
News International
Newspaper Publishers' Association
Newspaper Society
The Observer
Oak Industries
Office of Fair Trading
Open University
Ottawa Cablevision
Mr Simon Partridge
Pearson Longman
Philips Electronics and Associated Industries
Plaid Cymru
Polygram Leisure
Post Office Engineering Union
Project Thames
Racecourse Association
Radio Clyde
Radio, Electrical and Television Retailers' Association
Radio Telefis Eireann
Radio Tele Luxembourg
Rank Organisation

Red Rose Radio
Rediffusion
Richard Price Television Associates
Rockefeller Center TV
Rogers Cablesystems
Roman Catholic Church in England and Wales, Mass Media Commission
Royal Television Society
Saatchi and Saatchi Garland Compton
Samuels Jones Isaacson Page/BBDO
Satellite Television
Satellite Television Corporation, USA
Scottish National Party
Scottish Television
SelecTV
Mr Todd Slaughter, "Satellite TV News" Magazine
Society of Authors
Society of Cable Television Engineers
Society of Film Distributors
Society of Local Authority Chief Executives
Solent Cablevision
Sports Council
Standard Telephones and Cables
Mr John Starr
Swindon Viewpoint
Tavistock Gazette
"Technology Week" Magazine
Television South
Television South West
Test and County Cricket Board
Thames Television
J Walter Thompson Company
Thorn EMI
3Cs
Trades Union Congress
Trident Television
Tyburn Productions
Unda-Scotland
Union of Communication Workers
United Newspapers
Mr Stuart Varney, Cable News Network
Video Copyright Protection Society
Videomarketing
Videonet
Visionhire Cable
Visnews
Warburg Paribas Becker
Warner-Amex Satellite Entertainment Company
Welsh Fourth Channel Authority
Dr B J Witcher
Women's Broadcasting and Film Lobby
Women's National Commission
Writers' Guild of Great Britain
Yorkshire Television
Mr Alexander Zwissler